Jorge Alfredo González Pérez
Joao Custodio Ferrea

# Productive impact of different doses of organic matter in Beterraba

Jorge Alfredo González Pérez
Joao Custodio Ferrea

# Productive impact of different doses of organic matter in Beterraba

ScienciaScripts

**Imprint**



Cover image: www.ingimage.com

This book is a translation from the original published under ISBN 978-3-639-68415-5.

Publisher:
Sciencia Scripts
is a trademark of
Dodo Books Indian Ocean Ltd. and OmniScriptum S.R.L publishing group

120 High Road, East Finchley, London, N2 9ED, United Kingdom
Str. Armeneasca 28/1, office 1, Chisinau MD-2012, Republic of Moldova, Europe
Managing Directors: Ieva Konstantinova, Victoria Ursu
info@omniscriptum.com

Printed at: see last page
**ISBN: 978-620-8-56990-7**

## ACKNOWLEDGMENTS

- ✓ First of all, I would like to thank God for the life and health he has given me every day. I would also like to thank my parents for having given birth to me, for the effort and dedication they have always shown towards my education, and my dear mother, Rita, for her enormous sacrifice.
- ✓ To my siblings ã os for their unconditional strength and courage, for believing in me even when it didn't seem possible. My special thanks go to my colleague Raimundo "in memory" who was tireless, giving up his programs in order to give me some explanatory support until the wee hours throughout my training. To Dr. Justo Chicumbi for his strength and encouragement to see me graduate.
- ✓ To my tutor MSc. Jorge Alfredo González Pérez, thank you very much for your patience, dedication and zeal in transmitting your knowledge in a practical way, showing yourself available for new challenges, there were sleepless nights, always doing the impossible for the sake of this epic moment, to say that it was an honor to have been your tutor.

*"Every walk of 1000 km begins with the first step"*

## DEDICATORY

- ✓ I would like to dedicate this achievement to my dear mother Rita, who did everything she could to see this dream come true; if it weren't for her I wouldn't be here today,
- ✓ To all the people who, directly or indirectly, always gave that moral support to see this moment arrive.
- ✓ Dedication to all the teachers of the agricultural engineering course who, in the process of teaching and learning, transmitted practical knowledge.

*"To all of academic society I dedicate...."*

## SUMMARY

This work was carried out on the property belonging to farmer Mr. Henrique Mariano, located behind the SONANGOL fuel pumps, in the town of Homaengue, Kuanhama municipality, Ondjiva province. The sowing and harvesting season took place between the months of March 2023 and June 2023, the rainy season according to the National Institute of Meteorology and Geophysics (Annex 1), in which a diagnosis was made of the main problems that existed in the low yields of the Detroit Dark Rede variety of beet. For this purpose, percentages of organic matter from bovine manure of 25%, 50%, 75% and 100% were evaluated to assess which had the greatest influence on this crop; hence the percentage at 100% had the highest values with regard to bulb weight, but not so for foliage, bulb diameter and leaf length where this amendment had no effect. All the phytotechnical and phytosanitary work established by the Technical Manual for Organoponics, Intensive Gardens and Semi-protected Organoponics was carried out. The analysis was carried out using the Statistics program, through a completely randomized design and the Duncan test was used to compare the means.

**Key words:** Fertilization, treatment, mining, organic

**CONTENTS**

# I. INTRODUCTION

Agriculture is one of the main sources of development and progress for society, and soil is the most valuable resource on the planet, so it follows that its conservation and proper use is a major task for the agricultural sector.

Agriculture cannot be seen as a mining activity, which only extracts wealth from the soil without replenishing it, nor can it be an industrial activity that requires all inputs to be external. It must, to a large extent, produce its own inputs, maintain and restore the fertility of the soil, extract productive resources from it and return them duly enhanced and transformed without affecting the surroundings (Altieri, 1989; Hacht, 1991).

For thousands of years, agriculture was the most essential activity for the survival of mankind as it was the main source of food for man and raw materials for industry. The beginning of the development of agriculture was what determined the most transcendent changes of the period between 8000 and 5000 BC in the regions of Mesopotamia, Egypt and Persia. The author also adds that its discovery was probably due to women and not men, because while the latter were dedicated to hunting, women sowed the seeds of wild herbs (Garay, 1979).

The first forms of agriculture began with the cultivation of fruit and tubers in tropical areas (Riveiro, 1973). As agriculture became the basis of the population's diet, man took charge of it and gave hunting a secondary place (Garay, 1979). According to Núñez (1994), organic farming represents a return to agricultural methods before the industrial revolution, although it combines traditional methods of environmental conservation and biological balance with modern technology. This agriculture is a production system that excludes the use of synthetic compounds such as fertilizers, pesticides, growth regulators and food additives in livestock feed. In addition, the aim is to replace as much as possible external supplies, mainly industrial chemicals and fossil energy, with internal resources that can be obtained close to the farm.

Fertility is an integrative concept of the qualities of a soil that determine its ability to supply nutrients to plants, maintaining a high level of production that lasts over time. One of the basic principles of ecological agriculture, set out in Regulation EEC 2092/91, is to maintain and/or improve fertility and the soil's biological activity. The incorporation of crop residues, green manures, organic contributions and, possibly, inorganic fertilizers covered by the regulation are the most common practices.

The parameters that determine fertility can be grouped into physical, chemical and biological. In conventional agricultural systems, based on the use of inorganic fertilizers, the physical and chemical properties will be decisive in ensuring the availability of nutrients for the crops. Depending on these conditions, the most suitable fertilizers will be chosen for each situation and work will also be done to create good physical conditions. In ecological farming systems, where the dynamics of natural processes prevail, organic matter, in its different forms, is the most important factor, becoming the parameter par excellence that determines fertility (Lavrador, 1996).

It directly affects physical properties and microbial life establishes itself on it, activating biogeochemical cycles and releasing nutrients that can be made available to crops (Juma and McGill, 1986).

Organisms act on organic remains (fresh organic matter), taking advantage of part of their constituents and the energy associated with their molecules. The result of the organisms' activity is the formation of humic compounds which have a major impact on the soil's physical properties, while also providing a reserve of mineral elements. Similarly, the decomposition of organic matter releases mineral elements that form the basis of plant nutrition. The microbial population is therefore often a good indicator of soil fertility (Turco *et al.*, 1994).

The speed with which microbial populations change makes biological activity a very sensitive parameter for evaluating hypothetical changes in the soil environment (Dick, 1994).

Enzymatic activities seem to be useful parameters for measuring biological processes. There are numerous studies which show an increase in enzymatic activities when cultivation practices are applied with the contribution of organic compounds (crop residues, green manures, manure), while in systems based on monoculture or strictly mineral fertilization, a decrease is observed (Dick, 1994).

The use of enzymatic activities as a quality indicator is of great interest due to the apparent potential they offer. The main challenges are to standardize working methodologies and to have systematic studies in different environmental situations (Dick, 1994).

Parameters that indicate biological activity can be a very useful tool in agricultural systems aimed at maintaining or improving soil fertility through the management of organic matter. At present, studies carried out on Mediterranean ecological farming systems are still scarce (Albiach, 1997; Canet *et al.*, 2000).

Soil organic matter (SOM), made up of plant and animal residues in different states of decomposition, as well as microbial biomass, is closely related to the biological, physical and chemical properties of the soil (Christensen, 1996). In most soils, MOS is the main agent stimulating the formation and stabilization of aggregates, which are differentiated according to their size into macro-aggregates (> 250 μm) and micro-aggregates (≤ 250 μm). The incorporation of organic matter into soil aggregates protects it from rapid decomposition, determining its stability in the soil (Golchin et al., 1994).

On the contrary, soil cultivation encourages the decomposition of organic matter due to better aeration, which stimulates the activity of soil microorganisms. The accelerated development of an agriculture that is increasingly seeking the necessary pedestal of a column in its sustainable character and the human intelligence put into satisfying the ever-increasing food needs of the world's population have revitalized farmers' sense of productivity and maximum efficiency through the use of technologies and/or products that positively contribute to achieving this goal (Landa., 2002).

Nowadays, there is a worldwide trend towards the development of sustainable agriculture, where the aim is to minimize the use of chemical products, which are becoming more and more expensive, unbalance the environment and can also cause damage to animal and human health (López *et al.*, 2005).

Plants fertilized with organic sources are less attacked by insects, as they have a more adequate balance of nutrients. This was discovered by the scientist Chaboussov, who demonstrated the dependence between the nutritional quality of plants and the appearance of pests ACTAF, (2008).

In Angola and specifically in Cunene, with a few exceptions, there has been neither the culture nor the means to collect, take advantage of and apply the organic fertilizers that we have the potential to produce. Farming has been based on the application of chemical fertilizers, especially in intensive agriculture, which has affected the chemical and physical properties of the soil, reducing its capacity to produce crops and as a result has had social and economic repercussions for the country.

Organic fertilizers originate from plant and animal waste, which in its simplest form can be crop residues that remain in the fields and are incorporated spontaneously or with the cultivation work, and animal waste that remains in the field when the animals stay in pastures. Organic fertilizers that are too wet not only increase transportation costs, but are also difficult to manage, apply and distribute properly.

There is a basic principle demonstrated in various investigations that establishes a certain proportion of organic matter where the minimum value is 50 % to obtain high yields in a stable manner in crops, however, the quantities of each component in the mixture can be varied with good results, hence, in our particular case they are presented as:

**Problem:** Low yields in sugar beet cultivation in the Naipalala II neighborhood, Cunene Province.

**General objective:** To determine the influence and of different percentages of cattle manure on beet yields.

**Specific objectives:**

1. Determine the percentage of cattle manure that has the greatest influence on the cultivation of beet.
2. Determine the agro-productive behavior of the crop in the different treatments.

**Hypothesis:** If the percentage of cattle manure that has the greatest influence on beet yields is determined, higher yields can be obtained.

# II. LITERATURE REVIEW

## 2.1 General information about beet (*Beta vulgaris*, L). Origin

Beet descends from the wild species Beta marítima. L. (Becceerdilinge, 1950). According to Shomoth 1962, it is now widespread in a number of European countries along the Mediterranean coast, India and Egypt.

The botanical species Beta marítima, popularly known as "sea chard" or "wild chard", is a plant native to the coastal area of North Africa. Its cultivation is very old, dating back to the 2nd century BC. C., and gave rise to two different vegetables: one with abundant foliage, chard, and the other with a thick, fleshy root, beet. In principle, ancient civilizations only consumed the leaves of the beet. It is known that the Romans consumed this root, but it wasn't until the 16th century when it returned to the diet, in this case of the English and Germans (Guenkov, 1974).

Over the years, the cultivation of table beet has grown and improved. Today, its consumption is widespread in all temperate countries, especially in Europe, where France and Italy are the main producers. Although its production is not very high.

It was known to ancient peoples, but only the leaves were used as food. It was used as a medicinal plant and the dried roots as a substitute for coffee.

It is very energetic and advisable in cases of anemia, blood diseases and convalescence due to its high iron content. It is also rich in sugar, vitamin c, b1, b2, p and carotenes. The sugar contained in beet is sucrose (Guenkov, 1989).

Beet (*Beta vulgaris,* L.) is grown almost everywhere in the world, although on a more limited scale *than* carrots (Huerres and Caraballo, 1996). It is mainly harvested for fresh consumption, as a salad, due to its richness in sugars, mineral salts and carotine, substances of great importance for the vitality of the human organism in general. The leaves have great nutritional value, greater than that of the large, succulent roots, which are used in human food to extract sugar, depending on the characteristics of the different varieties and species.

It was introduced into agriculture around the fifteenth century, but it wasn't until the eighteenth and nineteenth centuries that the first selection work was carried out and valuable varieties were created that have been preserved to this day. it is grown all over the world, increasing its production and consumption (Guenkov, 1974).

### 2.2 Nutritional information

One cooked and rotated beet contains several nutritional elements (University of Illinois Extension, 2009) (table 1)

**Table 1.** Nutritional content of a rate of cooked and roasted beet.

| # | Elements | Contents |
|---|---|---|
| 1 | Heat | 31 |
| 2 | Proteins | 1.5 g |
| 3 | Carbohydrates | 8.5 g |
| 4 | Dietary fibers | 1.5 g |
| 5 | Potassium | 259 Mg |
| 6 | Phosphorus | 32 Mg |

**Health**

Studies carried out in 2008 indicate that drinking half a liter of beet juice a day reduces hypertension, i.e. when mixed with saliva it turns into nitrite, which is transformed by hydrochloric acid in the stomach into nitric oxide, which in turn acts to reduce hypertension (Beta vulgaris, 2009).

**Beet leaves**

An ingredient in the most popular soups, they have a less pronounced flavor than spinach and less iron, but they do contain vitamins A, B1, B2, B3 and C. They are soothing, laxative, diuretic and refreshing.

### 2.3 Economic importance and geographical distribution

Biological properties offer the possibility of harvesting more than one crop in a year from the same field. They allow not only an increase in productivity of the land, but also economic use of the resources imposed to improve the soil (Guencov, 1974).

It is mainly produced for fresh consumption, in specialized areas of companies in the different provinces of the country, school gardens and cooperatives, but always in small areas. Beet sugar is currently grown at three times the rate it was five years ago, and in absolute production figures it has overtaken sugar cane; this is due to the modernization of cultivation as well as the decrease in fodder beet production (Infoagro, 2009).

### 2.4 Food importance

Almost all types of vegetables are relatively lacking in energy nutrients: carbohydrates, albumin, fats. Therefore, human beings would not be able to satisfy their food needs using only vegetables. The value and indispensability of vegetables for their nutritional quality lies in their richness in vitamins, easily assimilated organic acids, minerals and essential oils. These substances play an exceptional role in the development and normal functions of the human organism. They help to improve the taste of food, increase the secretion of the digestive glands, and with it the better digestion and assimilation of other nutritional substances.

Vegetables are very important for regulating the action of the nervous system and for increasing the body's resistance to different diseases. Their greatest importance from a dietary point of view is fundamentally due to their iron and phosphorus content, with a much lower content of other elements, both minerals and vitamins (Guencov, 1974).

### 2.5 Taxonomy and morphology

Beta vulgaris L., the beet, also known as white chard, betarava, betarraga, beterava, beterraga, and betabel, is a plant from the Chenopodiaceae family, of which the leaves and root are edible.

The table variety has thick, red and fleshy roots, which are mainly eaten cooked; the color is due to two pigments, betacyanin and betaxanthin, which are indigestible and dye food, excrement and urine that color. However, due to its non-toxicity, it is often used as a coloring agent in food products (*Beta vulgaris*, 2009) (Table 2).

Beet or betabel is the deep, large and fleshy root that grows on the plant of the same name. It belongs to the Chuenopodiaceae family, which includes 1400 species or plants, almost all of which are herbaceous and native to coastal areas or temperate saline soils. This family also includes other vegetables as popular and nutritious as spinach and chard (Infoagro, 2009).

**Table 2.** Taxonomy and morphology of beet (*Beta vulgaris* L.)

| | |
|---|---|
| Kingdom: | Plantae |
| Division | Spermatophyta |

| | |
|---|---|
| Subdivision: | Magnoliophytina |
| Class: | Magnoliatae |
| Order: | Cariophyllales |
| Family: | Chuenopodiaceae |
| Gender: | Beta. |
| Species: | Beta vulgaris |

**Root:** It has a developed and highly branched root system. The main root extends up to 1.50 - 2.0 m, while the secondary roots extend laterally up to 60 cm. Given its characteristics, it shows a certain resistance to drought. It is almost completely buried, with a greenish-yellow skin that is rough to the touch, and is the most important part of the reserve accumulator (Infoagro, 2009).

**Stem:** The stem has limited growth in the first year, being located at the point where the fleshy root and the leaves intersect. The floral stem grows after the fleshy root has emerged. The branched floral stem can reach a height of 0.80 - 1.20 M. Each branch ends in a flower (Guencov, 1974).

**Leaves:** The leaves are simple and arranged in a rosette. The limb is triangular and green or purplish in color. The peduncle is broad and somewhat fuzzy in some varieties (Guencov, 1974).

**Flowers:** Not very showy and hermaphrodite. Fertilization is usually crossed because their reproductive organs mature at different times. They have 5 sepals and 5 green petals with reddish pigmentation (Guencov, 1974).

**Seeds:** In the process of fertilization, several flowers are welded together, giving rise to glomerules which are generally called seeds, but in reality each glomerule is made up of 2 - 4 seeds. They are attached to the calyx and are woody (Guencov, 1974).

The beet seed is actually a glomerulus made up of several seeds enclosed in the same suberous covering. It has the disadvantage that several plants are born at the same point, making the work of clarification more difficult and expensive.

## 2.6 Edaphoclimatic requirements

Ecological factors are of great importance for plant growth and development. This has to do in one way or another with the physiological processes that take place inside and outside the plant: photosynthesis, respiration, assimilation, de-assimilation of substances, growth and development.

**Soils:** Beetroot can be sown on different soils, as long as they provide an adequate balance of nutrients and moisture. The most suitable are deep soils with good texture and absorption capacity, permeable soils and high moisture content (Guencov, 1974).

Deep soils with a pH of around 7, high water tension capacity, low crust-forming density and good aeration are the most suitable for sugar beet (Infoagro, 2009).

Growth is adequate when the pH is 6 - 7. In acidic soils with a pH below 5 or alkaline above 8, the plants show different physiological disorders, the seedlings sprout slowly and a large percentage of them perish (Guencov, 1974).

**Transplant**

By transplanting, the growing cycle is advanced, thus bringing forward the sugar beet harvesting season. Transplanting also offers major agronomic advantages, such as:

a) Increase crop yield by up to 25%.
b) Seed cost reduction of up to 58 %.
c) You avoid the problem of hatching, as well as that of re-seeding.
d) Facilitates the fight against bad weeds and saves on the use of herbicide treatments.
e) It saves on the supply of insecticides for possible infestations in the crop, as transplanting facilitates the fight against certain pests (Infoagro, 2009).

**Soil moisture**

Beet is a demanding species when it comes to soil moisture. It is most demanding during the germination phase and in the early stages of growth. Once the root system has formed, the demand for moisture decreases. However, for good production of fleshy roots, normal and constant soil moisture must be maintained so that growth and development are uninterrupted. It does not tolerate excess moisture in the soil (Guencov, 1974).

It is a plant demanding of soil humidity with a percentage that must be between 60 and 70 °C. When it is low, a high percentage of stomata close, transpiration and self-cooling of the meshes is reduced and the entry of $CO_2$ into the photosynthetic process is affected, reducing the accumulation of carbohydrates (García, 2007).

**Watering:** From sowing to 8 days after germination, 2 times daily, number of waterings in the 10 - 20 stage, watering time for microjet 14 minutes, watering standard L/$m^2$ is 4. From 8 days to 60 days watering is daily, the number of waterings in the stage is 52, watering time for microjet 17 minutes, watering norm L/$m^2$ is 5. From 60 to 90 days, watering alternates, the number of waterings per stage is 15, watering time for microjet 14 minutes, watering standard L/$m^2$ is 4, (Technical Manual for Organoponics, Intensive Horta and Semi-Protected Crops and IIRD, 2007).

Water is the factor that most influences the weight and richness of sugar beet; it is also the most difficult to control, as it depends on many other parameters such as climate, soil type, root depth, etc. The volume of water to be used can vary between 50 and 70 L/$m^2$, being applied from mid-August to early September. Beet needs approximately 20 L/$m^2$ to sprout, but if it hasn't received water again within 15 - 20 days, it could lose its seed (Infoagro, 2009). According to Gómez *et al.* (2000), the pH should be between 6 - 7.5. The concentration of chloride ions is equal to 0.009 moles/liter.

**Climate:** one of the main factors directly affecting yields. A temperate, sunny and humid climate contributes to the production of a high percentage of sugars in beet. Illumination is very important in this crop, as it allows photosynthesis to take place properly and conditions the importance of sugar production (Infoagro, 2009).

Air humidity: This plant is demanding of air humidity, which is why sprinkler irrigation is more favorable due to the cooling it produces to the leaves, reducing transpiration. This plant needs 20 L of water/$m^2$ to germinate, but if it hasn't received water within 15 - 20 days, it may not germinate.

**Temperature:** the thermal balance is of exceptional importance for the propagation of horticultural plants. Photosynthesis, respiration, enzymatic activity in the cells, their division and growth, the absorption capacity of the roots and other processes depend on it. Through photosynthesis, nutrients extracted by the roots and $CO_2$ taken from the air by light, plants synthesize complex nutrients for their sustenance and development.

At the same time as photosynthesis, respiration takes place, which breaks down the complex organic substances synthesized through photosynthesis and releases the energy needed for the vital processes of plants (Guencov, 1974).

**Light:** The importance of light for the life of green plants is generally known. Without light there is no photosynthesis, therefore no concentration of organic matter, no growth and development, no life. If this factor is limited, the productivity of photosynthesis is reduced, as is the quantity of organic substances synthesized and, consequently, the quality and quantity of production (Guencov, 1974).

**Day length:** The importance of day length lies in the fact that the various species and varieties of vegetables were formed in geographical regions with different day lengths, i.e. different proportions of dark and light periods during the 24 hours of the day (Guencov, 1974).

**Nutrition: this is** of great importance because plants must not suffer any shortage of nutritional substances in order to form organs and carry out all biological or physiological processes normally (Guencov, 1974).

A shortage of one or other elements, or an abnormal proportion between them, has an unfavorable effect on the growth and formation of organs. In such conditions, harvests decrease or the quality of production worsens.

### 2.7 Nutrition. Nutrient requirements and extraction

### Cultivation requirements.

Beet's water needs are considerable. Beet's leaf surface can be considered one of the most developed of all the different crops. As transpiration takes place through the leaves, the plant expels very significant amounts of water which it must first take up from the soil. It is estimated that to produce 40 tons of roots, the crop can evaporate 7,000 cubic meters of water per year, which is equivalent to 700 L/m of rainwater[2].

In the cultivation of beet, the intensity of lighting is very important, as it allows the chlorophyll function to be properly carried out and conditions the importance of sugar production (Infoagro, 2002). Any substance that is added to the soil to contribute one or more nutrients to the plant in order to increase its growth is a fertilizer (Cooke, 1987).

A study carried out by INSAM, (1996), shows that there is a clear analogy with the effects on plant development of the application of humic substances on the mineral nutrition of plants, effects on the accumulation of phosphorus, nitrogen, potassium, calcium and magnesium. It was determined that for the production of 1t, 2.7 Kg of N, 1.53 Kg of $P_2O_5$ and 4.31 Kg of $K_2O$ are extracted.

**Nitrogen:** Adequate amounts of this element are fundamental in the formation of beet foliage. Deficiency causes the plant's leaves to be small and pale in color, usually yellowish, with reddish tints, with symptoms beginning with the older leaves. When the deficiency is acute, the harvest is greatly reduced. Nitrogen in the form of nitrate is unfavorable for the plant (García, 2007).

Excess nitrogen increases leaf development, but decreases the ability to mobilize sugars to the root, which is essential for the foliage. A lack of this nutrient causes small, pale leaves, usually yellowish with a reddish tint, with the symptoms beginning with the older leaves, which reduces the harvest (Infoagro, 2009).

**Phosphorus:** With a deficiency, the plants grow stunted, the leaves take on a dull green color, sometimes slightly tanned and with purplish spots. Young plants assimilate phosphorus more intensely than old ones. This rapid absorption of phosphorus by young plants is related to their rapid development during the period of maximum growth (García 2007).

The effectiveness of phosphorus is manifested mainly in the young stages of the plant, so it is recommended to bury this element as early as possible so that it is available and assimilable in the early stages of the beet (infoagro, 2009).

**Potassium:** An adequate supply contributes to the sugar content of the plant

and increases yields. Deficiency manifests itself in marginal and winter burns, followed by wilting and collapse in the older leaves. It interferes in the formation and transportation of carbohydrates from one plant organ to another (Guencov, 1974).

Dinchev, (1977), stated that potassium belongs to the group of elements that are unconditionally necessary for plants, most of the potassium in plants is found in the cell juice and can be extracted easily with water.

Boron: This is one of the most important micro-elements. Normally, 20 kg/ha of borax, distributed with the fertilizer before sowing, is sufficient. The drawback is to achieve uniform distribution, but combinations with boron, such as boron superphosphate, can be used (Guencov, 1974).

Calcium: One of the main symptoms of calcium deficiency is intervernal chlorosis in older leaves (starting at the apex and edge) followed by necrosis (Guencov, 1974). It plays a role in the formation of cell walls and in the adhesion of one cell to another, through its combination with peptic acids in the formation of calcium pectinate.

Magnesium: Magnesium deficiency is visible in yellow spots on the leaves and often occurs in light soils. It is recommended to spray with liquid fertilizers containing magnesium (infoagro, 2009).

**Microelements:** They are of great importance because it is an established fact that on several occasions, especially with regard to some types of soil, small amounts of boron, zinc, copper, manganese, magnesium and molybdenum greatly favor the growth and development of vegetable plants and, of course, yields. The importance of these elements is much greater during the early stages of plant development . This is why the use of solutions of these elements on seeds or young plants has often been very useful (Guencov, 1974).

**Organic** fertilizers**:** Beetroot's nutritional requirements are high and fertilization must take into account its long vegetative cycle, which requires sources that are available and assimilated quickly, on the one hand, and long-acting and resistant nutrients, on the other. Soils that tend to become compacted should be fertilized with organic products to improve their structure. 22,000 kg/ha of manure is recommended (Arzola P. N. *et aL.*, 1981).

Manure is directly and indirectly important for the growth and development of vegetable plants. It improves the texture of the soil and contributes to good air penetration, moisture absorption and gas exchange. In soils that are richer in organic substances, microbiological action is more active and the number of microorganisms is greater. All of this creates conditions for better growth and development of the root system and, as a result, more intensive extraction of nutrients.

Organic fertilizer improves the physical conditions of both heavy and light soils. The former become looser and more airy and the latter more compact and able to retain moisture. Manure contains all the elements that vegetable plants need. The nutritional elements are released little by little in parallel with the decomposition of the complex organic combinations, so this organic fertilizer provides the plants with nutritional substances over a longer period of time. For this reason, manure is very convenient, particularly for species and classes of plants with a longer growing cycle.

Fresh manure, especially when it is mixed with foreign substances (straw, sawdust, etc.), is not very suitable as a fertilizer. Instead of enriching the manure I am used to, it can deplete it of nitrogen, and therefore its use is not preferable (table 3).

**Table 3.** Characteristics of the organic fertilizers most frequently used in agriculture (data in fresh support)

| TYPE OF ORGANIC ALLOWANCE | ANALYSES | | | | | |
|---|---|---|---|---|---|---|
| | HUMIDITY | RELATION SHIP | M.O. | N. | $P_2O_5$ | $K_2O$ |
| | % | C/N | % | % | % | % |
| Bovine manure | 80.00 | 20:1 | 11.50 | 0.33 | 0.23 | 0.72 |
| manure | 67.40 | 30:1 | 17.93 | 0.34 | 0.13 | 0.35 |
| Pig manure | 72.80 | 19:1 | 15.00 | 0.45 | 0.20 | 0.60 |
| Sheep manure | 61.60 | 15:1 | 21.12 | 0.82 | 0.21 | 0.84 |
| Compound | 75.00 | 16:1 | 13.75 | 0.50 | 0.26 | 0.53 |
| Chicken | 75.00 | 22:1 | 15.54 | 0.70 | 1.03 | 0.49 |
| Bat dung | 23.00 | 8:1 | 13.20 | 0.96 | 12.00 | 0.40 |
| Peat | 70.00 | 42:1 | 14.40 | 0.20 | 0.17 | 0.12 |
| Cool slowness | 71.00 | 30:1 | 16.40 | 0.32 | 0.60 | 0.17 |
| Slowness cured | 54.50 | 15:1 | 28.90 | 1.11 | 1.11 | 0.15 |

Note: The table expresses average values that can serve as a reference for evaluating organic subsidies, but should not be taken as definitive, as they may vary according to their origin. Each producer should have a characterization of the organic subsidy they spot (Collings, 1958, cited by Gros, 1966; Pequeno 1966 and Fundora et. al1983).

**Cultivation characteristics**

The beet (Detroit Dark Rede) has a perfect globe-shaped bulb of medium length. The leaves are reddish green. Its biological cycle is 80 to 90 days and its sowing date is between September and May. In organoponic conditions and intensive vegetable gardens, it can be sown all year round. Good results can be obtained by transplanting in good time and increasing the frequency of watering during periods of high temperatures. According to the technical manual for organoponics, intensive vegetable gardens and semi-protected dagger organ (2007).

**Preparaç ã o do solo:** In order to get a good yield this crop, it is necessary to drill 35 - 45 cm into the **soil** to facilitate good root development and conserve as much water as possible (Infoagro, 2009).

**Sowing season**

Normal: September - May.

Optimal: October - January.

**Sowing:** The most suitable sowing period in the country is November-December, but in order to achieve greater distribution to the population over a longer period, sowing begins in September-February (García 2007).

**Thinning:** the best plants are selected. It is almost always done manually and it is necessary to remove the focus plants so that they are approximately 7 - 15 cm away. If the removal is delayed until the plants are 7.5 cm tall, the plants can be used like spinach (García 2007).

**Transplanting:** This technique consists of obtaining strong plantations, which would result in agronomic advantages such as:

- ✓ Increase crop yields by up to 25%.
- ✓ Seed cost reduction of up to 85 %.
- ✓ Avoid re-seeding.
- ✓ Facilitates the fight against non-crop plants.
- ✓ Facilitates the fight against pests and diseases.

**Fertilization:** Betarraga is an undemanding crop, so it adapts to a wide range of soils, provided they have a good organic matter content and are relatively free of nematodes. it is recommended to apply manure at a rate of 20 MT/ha/year during land preparation. If the previous crop received a medium or high dose of manure and fertilizers, it is usually not necessary to re-fertilize the plant with synthetic fertilizers. If it were used, it would only be nitrogen (N) sources in low doses and with the fertilizer applied in a line to the bottom of the freshly moistened furrow (Guenkov, 1974).

Fertilizing the land dedicated to ecological agriculture is one of the pillars of this form of cultivation. It is very practical to produce your own fertilizer, compost production is one of the most widely used. In ecological agriculture, the aim is not to directly nourish the plant, but to stimulate the whole, i.e. the soil and the plant, maintaining or improving the fertility of the soil by "favoring the humic archaic complex and the development of soil microorganisms".

Organic matter is the basis of fertilization, although you can also use green manure as fertilizer, which consists of growing and burying a plant so that when it decomposes it turns into manure, especially using legumes. These enrich the soil especially in nitrogen thanks to bacteria that live in their roots and fix atmospheric nitrogen, which the plant gives up to the soil in the form of manure when it is buried. Contrary to popular belief, keeping the soil covered helps conserve it by improving the use of water and nutrients. Live plant covers, mulching, etc. can be used. The mineral fertilizers that can be used are from natural sources that have been extracted by physical processes (A. Ecológica, 2009).

**care:** It is often important when they are small, because they grow very close to the surface, so frequent and superficial care is effective in eliminating plants that are not the object of cultivation (García, 2007).

**Harvesting: this** takes place when the fleshy root has reached a size suitable for consumption, since if it is allowed to increase in size and cycle, the fruit will be fibrous (García 2007).

**Growing requirements:** Beetroot is well suited to saline soils, and it even benefits from sodium, which it absorbs in abundance. In some parts of England, some sodium chloride has been applied as a top dressing. These contributions tend to underestimate the plant, but also to substitute Na in part for K in sugar production, since sodium chloride is cheaper than potassium chloride. In any case, the contribution of common salt to clay soils is dangerous because it worsens the physical conditions of the soil and favors the formation of crusts.

As a general rule, it is not advisable to repeat the cultivation of beet on beet in order to reduce the problems of pests and diseases, and also to avoid the growth of bad weeds that are poorly controlled with the herbicides used on the beet.

## 2.8 Main infestations and diseases

**Pests:**

- *Leaf miner* (Lariomyza sp): Attacks in the larval stage by opening a gallery inside the leaves.
- *Aphids* (Myzus Persicae.Sulzer): Forms colonies on the underside of leaves, sucking sap from the plant.

**Diseases**

- *Mushroom* (Cercospora Beticola Sacc): Causes circular spots of 2 mm and edges with a more intense color, causing it to wither and die.

**Pest and disease control**

Ecological cultivation must be based on preventive methods, enhancing the good development of plants and, therefore, their natural resistance to pests and diseases. Prevention must be maximized through appropriate cultural practices that ensure the plants develop well and are therefore more resistant. Native species and adequate watering make plants more resistant.

By avoiding the cultivation of a single species, diversifying the species planted makes it more difficult for pests to appear, using appropriate rotation and association in the fields. The soil must be balanced to obtain strong plants and varieties adapted to the area will be used.

It is advisable to promote the development of native auxiliary fauna by using hedges and releasing useful insects (parasites and/or predators), such as Aphidius colemani. Ultimately, different products of natural origin can be used, such as pyrethrum, which is obtained from dried chrysanthemum flowers, or Bacillus thuringiensis, which is an aerobic bacteria that only attacks harmful insects.

**Pheromones, attractants and repellents**

Garlic extract is biodegradable and is used to repel whitefly, birds and different types of suckers. It is based on a food scent masking agent, pheromones (prevents pests from reproducing) and disconcerts birds because garlic is irritating to birds. It doesn't prevent this method from being ineffective for birds during periods of high hunger; other methods can be used, such as ultrasound or gas explosions with motion detectors. Garlic extract can mask the scent of some pheromone traps and can make them more ineffective (Revista Agricultura Ecológica, 2009).

**Weeds**

The presence of weeds in sugar beet cultivation is primitive in both technical and economic terms; technically because of the difficulty of controlling weeds, and economically because of the repercussions on production costs and the final gross product, whether using manual or mechanical weeding or the application of herbicides (Infoagro, 2009).

**Ground maintenance**

**Biological**

It naturally has a multitude of living organisms that carry out continuous "work": the roots as they explore in search of water and nutrients; roundworms, insects and rodents, with their galleries; other organisms with their exudations and residues that help to bind the particles of clay and humus together. Nor should we ignore the large amount of organic matter contributed by these organisms, as well as the conversion of organic matter into material that can be assimilated by plants. Various experiments have shown that biological work has advantages over mechanical work. These include

a) It doesn't condense the soil as it passes through it, which is usually the case when working the land with machinery and which makes it necessary to work deeper every so often.
b) Mowing the grass and leaving it as mulch produces several improvements: on the one hand, the sun doesn't dry out the soil, preserving moisture, and on the other hand, it serves as a shelter for microorganisms and other organisms.
c) On the other hand, adventitious plants, also known as "bad weeds", serve as hosts for useful insects, absorb nitrogen that would otherwise be lost when it evaporates into the atmosphere and then return to the soil when it is converted into compost. And if the plant has deep roots like alfalfa, then it extracts nutrients that are otherwise inaccessible to other plants with root systems that are less capable of going deep. To prevent the field from filling up with adventitious plants, crop rotations are carried out with false sowings and balanced tillage programs.

**Mechanic**

The main condition an apero must meet is not to turn the soil over too deeply so as not to alter the natural order of the soil, working with seasoning and not abusing it, thus partly avoiding the undesirable effects of mechanical work such as soil mineralization and compaction due to the weight of the machinery (Revista Agricultura Ecológica, 2009).

For practical and historical reasons, the reasons for using mechanical work are:

a) Faster work in the field, both sowing, harvesting and treatments.
b) Eliminate the competence caused by adventitious plants.
c) The transportation of produce from the same field to its destination.

### 2.9 Organic allowances

Organic fertilizers include: animal manure, crop residues, green manure, industrial waste, compost, worm humus and, more recently, biofertilizers or microbial inoculants, which can be included among them due to their characteristics.

**For short-cycle crops.**

When applying organic fertilizers to soils dedicated to short-cycle crops, the recommended quantities are low Formation of organic matter in soils. Soil receives a large amount of organic waste from different sources. These include the remains of higher plants, agricultural crops and, to a lesser extent, animal waste. They reach the soil in two ways: they are deposited on the surface (leaves, branches, flowers) or they are directly caught up in the soil mass (roots).

A layer of organic remains known as humus accumulates on the surface of forest soils. In temperate climates, this horizon can have between 10 and 70 $t.ha^{-1}$ over time, despite receiving around 4 $t.ha^{-1}$ of plant remains each year. Tropical forest soils, on the other hand, receive between 100 and 250 $t.ha^{-1}$ a year and lack this organic horizon, which in any case can reach only 10 t.ha-1. This is due to the fact that in the tropics the action of the fauna and microflora of the soil is much more energetic and develops with greater activity throughout the year.

The desirable levels of organic matter in cultivated soils vary from in arid areas to and more in fertile valleys. Scientists point out that it takes time to maintain or increase the content of organic matter in soils that are intensively cultivated. In addition, they estimate that spraying 25 tons of fertilizer per acre (61 $t.ha^{-1}$) would take approximately 20 years to build up the soil's organic matter by 1% (INIFAT, 2002).

The production of compost requires the establishment of a strict temperature and humidity regime that guarantees the fermentation process and favors the decomposition of organic waste and the production of humus.

**- Roundworm humus**

It is produced by the decomposition of organic waste by specialized worms that are able to produce high-quality humus.

**Richness of roundworm humus**

Roundworm humus is five times richer in nitrates, twice as rich in calcium, 2.5 times richer in magnesium, seven times richer in phosphorus and eleven times richer in potassium than humus from high-quality soil. High-quality soil generally has 150 - 200 million microorganisms per gram, while worm humus has 250 - 300 million diverse microorganisms per gram that are beneficial to the plant (Pinheiral, 1994). (table 4)

**Table 4.** Analysis of roundworm humus

| | |
|---|---|
| | 30-66 % |
| pH | 5.6-7.9 % |
| matter | 35-7 |
| | 15-68 % |
| N | 1.4-3.0 % |
| P2O5 | 0.2-5.0 % |
| K2O | 0.2-2.5 % |
| Ca | 2-12 % |
| Mg | 0.2-2.6 % |
| | 0.6-0.9 % |
| | 66-1467 ppm |
| Cu | 34-490 ppm |
| | 87-1600 ppm |
| B | 26-89 ppm |
| Co | 9-48 ppm |
| Microbial Carba | 5 x 10-2 x 1012 |

**Source:** Pineda, 1994

**Importance of roundworm humus**

Worm humus has two elements that are very important for the plant: acidity and bacterial flora. Humus is a neutral substance, so the value of worm humus is excellent, as it is very close to the data obtained only from the best organic fertilizers.

The bacterial flora in this organic fertilizer reaches 2 trillion bacterial colonies per gram of fertilizer, instead of the few hundred million present in the same amount of fermented animal manure, which is considered one of the best.

A matter of indisputable practical importance is that roundworm humus, although given in excessive doses, does not burn any plant, not even the most tender (Ferruzzi, 1987).

**Main effects of roundworm humus**

The action of roundworm humus improves the structure of the soil because it acts as a cementing agent between the soil particles, giving rise to granular structures, which allow the soil to be formed:

a) Improving radical development
b) Improve gas exchange
c) Activating microorganisms
d) Increase the oxidation of organic matter and therefore deliver nutrients in chemical forms that plants can assimilate.
e) Use in any dose, without burning or harming the most delicate plants, as its pH is neutral.
f) Provide microelements in different proportions
g) Administer enzymes, which continue to break down organic matter, even after it has been expelled from the roundworm's digestive tract; these enzymes are classified as proteases, amylases, lipases, celluloses and chitins.
h) Use as a foliar fertilizer, due to its water-soluble nutrient components (Ferruzzi, 1987).

**Main characteristics of roundworm humus**

Roundworm humus improves the soil's physical characteristics and maintains the soil due to its colloidal structure, as it increases water retention capacity. It is a fertilizer that slowly releases its nutrients, is rich in trace elements and contains humic and fulminic acids that prevent the formation of mushrooms and mycetes (Ocsa, 1995).

## 2.10 Other organic allowances

### Bat dung

It is produced in pits by the accumulation of droppings and bodies of these mammals where they live and which sometimes constitute large reserves of organic matter rich in phosphorus.

### Compound

The chemical, physical and biological characteristics depend on the nature of the waste used to obtain or prepare it and the technological process employed. If cattle manure or other animal waste is used in its preparation, the compost will have a high humus and nitrogen content and a low C/N ratio. If vegetable waste with a predominance of grass or peat species is used in the preparation of the compost, the compost will have a low N content and a high C/N ratio and will generally have poor chemical and physical quality (table 5).

**Table 5.** Chemical analysis of the elements in the compound (data in fresh support), data set as INCA, 1998

| | |
|---|---|
| Humidity | 70 % |
| pH | 6.8 |
| Organics | 13.80 % |
| Total nitrogen (N) | 0.50 % |
| Total phosphorus ($P_2O_5$) | 0.26 % |
| Total potassium ($K_2O$) | 0.53 % |
| C/N ratio | 16:1 |

**Peat**

Peat is an accumulation and deposit of organic matter produced in areas where the accumulation and permanence of water on the surface of the soil for a long time limits microbial activity, leading to the accumulation of organic matter in large quantities. The quality and nutrient content of peat depends on the nature of the organic waste and its degree of decomposition.

**Slowness**

Liqueur is obtained as a result of the process of clarifying barrel juices in the sugar industry, by alkalizing with CA(OH)2 and applying heat, coagulating and precipitating the solids in the juice and then separating them by decantation and filtration. Lentil is an organic fertilizer rich in organic matter, phosphorus and calcium. The production of sludge is equivalent to 3-4 % of the weight of the barrel that processes the plant.

**manure**

gallinaza is the name given to the organic fertilizer obtained from excreta and other waste in places where poultry is intensively reared for the production of eggs and meat.

This organic fertilizer in its fresh state contains many substances that are in the process of decomposition and when applied produce changes in the soil and affect plants; for this reason it is necessary that before using them, they are fermented and decomposed (INCA, 1998).

**Bovine manure**

**Methods and forms of application**

When organic fertilizers are used to improve the physical, chemical and biological characteristics of the soil, they need to be sprayed all over the land and incorporated into the soil preparation work. The most suitable method of application is to use an organic fertilizer spreader.

The best time is to apply it after the first pass of the plow and incorporate it into the soil with the second pass of the plow. Every producer who uses organic fertilizers must bear in mind that in order to obtain the maximum benefits from these investments, it is essential to incorporate them into the soil. Manure should be applied in the furrow before sowing and should be incorporated into the soil with the seed (INCA, 1998).

Alonso, (1980), points out that the importance of manure is not only due to the amount of organic matter it contains, but also to the nutritional principles it imparts to young plants. Its effects are especially noticeable in clay soils, as well as calcareous and sandy soils in which there are favorable changes in cohesion, as it lowers their already low pH, which is detrimental to vegetation, which is why he refers to its contribution to food.50 of nitrogen, 0.25 of phosphoric acid and 0.50 of potassium, figures that speak for themselves of the importance of this material.

López and Zada (1989) point out that manure is an amendment capable of contributing mineral elements. As an amendment, it contributes the organic matter necessary for the nutrition of soil microorganisms, in line with Pérez et al. (1968), who state that the value of manure depends on the nutrient content of the plants and its effectiveness as a soil conserving and building agent, exerting a great effect on soil microorganisms, as well as its structure.

Gorinij, (1961) and Jaivebo, (1967), agree that mineral fertilizer causes a greater increase in the harvest than manure, only during the first few years, but that in successive years with manure the harvests are higher.

It has been shown that the use of excreta when watering potreros and/or forage areas gives the soil and plants back some of the nutrients they have lost, and avoids water contamination. In addition, in cow pens, producing biogas with excreta to light and/or cook and then using the biogas plant's effluent as an allowance is even better (Redino and Pedraza, 2002).

### 2.11 Physical and chemical characteristics of organic fertilizers

The characteristics of organic soils are governed by their organic matter content, the nature of the materials involved in their formation and the fermentation and decomposition process to which the organic waste has been subjected.

The quality indicators of an organic allowance are set by:

**- Its humus content and quality**

Humus from the decomposition of organic waste with a low C/N ratio and low lignin content is looser and more friable, and its microbial flora is more favorable for the physical and biological conditions of the soils where it is applied, than when the organic fertilizer is obtained from plant waste rich in lignin with a very high C/N ratio. An organic fertilizer must have 50% or more organic matter in dry matter. Below 50% organic matter is considered to be of poor quality (INCA, 1998).

**- Its nutrient content**

Organic fertilizers must have a balanced N, P, K, CA, Mg nutrient content, so that when they are used they improve soil fertility and benefit the nutritional status of plants. In addition, organic fertilizers should not contain substances that acidify or alkalize the soil and could affect the normal development of crops (INCA, 1998).

**-Your moisture content**

Moisture content is the element that most limits the use of organic fertilizers due to the cost involved in transporting and handling large quantities of water in the organic fertilizer. In general, organic fertilizers have a moisture content ofbetween 60% and 80%, depending on the process used to obtain them and the nature of the organic waste they contain, which always tends to retain water. However, anyone who produces and manages organic fertilizers should ensure that they have approximately 60% moisture or less when they are used (INCA, 1998).

Organic fertilizers that are too wet not only increase transportation costs, but are also difficult to digest, apply and distribute properly. Indicators of the value of some species' excrement (tables 6 and 7)

**Table 6.** Excreta production and nutrient content in different species.

| Species. | Excreta t/year/500 kg | M.S. % | Nutrient content of one ton of fresh excreta in kg. | | |
|---|---|---|---|---|---|
| | | | N | $P_2O_5$ | $K_2O$ |

| Beef cattle | 9.5 | 20.0 | 6.3 | 4.1 | 4.8 |
|---|---|---|---|---|---|
| Dairy cow | 13.4 | 21.0 | 6.9 | 2.1 | 5.3 |
| bovine | 6.7 | 35.0 | 12.5 | 4.2 | 10.7 |
| Pork | 17.9 | 25.0 | 4.5 | 2.9 | 4.0 |
| Horse | 8.9 | 40.0 | 6.1 | 2.1 | 6.4 |
| Birds | 5.0 | 46.0 | 13.9 | 8.2 | 3.7 |

**Table 7.** Contributions of N, P, K, per day by cattle manure and urination.

| Category | N/g/animal/day | $P_2O_5$ | $K_2O$ |
|---|---|---|---|
| Cow | 118 | 20.8 | 89.0 |
| heifer | 77 | 15.4 | 65.0 |
| Taurus | 84 | 14.0 | 69.0 |
| Torete | 38 | 9.8 | 38.0 |
| Calf | 31 | 5.6 | 31.0 |

When animals are confined to pasture with partial stabling, the quantities of excreta that can be collected are much smaller. Poor handling of excreta or its exposure to the environment can lead to major nutrient losses.

**The importance of organic matter in the soil**

The development of crops depends on the soil's ability to provide the necessary quantities of nutrients for their proper development. The availability of these nutrients depends on various factors, with the content and quality of the organic matter present being one of the most decisive.

A fertile soil must necessarily have an adequate organic matter content, which ranges from 2% for sandy soils to 6% for moist soils.

The favorable influence of organic matter and especially humus on soils has been recognized since ancient times.

Its main effects on the chemical, physical and biological properties of the soil stand out (INIFAT, 2002).

**Functions of organic matter in the soil**

The functions of organic matter in the soil govern important processes in the soil-plant system and affect its properties. Let's highlight some of the most important ones.

**Physical properties:**

a) Aggregate formation and structural stability (García, 1992). Humic substances, like the polysaccharides present in organic matter, play an important role in the stability of soil aggregates (Swift, 1991). And therefore in maintaining their structure.
b) Favoring water penetration and retention (Waters and Oades, 1991).
c) Reduced erosion (Gabrielis and Michiels, 1991).

**Chemical properties:**

a) Maintains and increases the content of organic matter in the soil.
b) It increases the capacity of plant life to change and reserve nutrients (Cegarra et al, 1983).
c) Increasing the cushion capacity of soils, avoiding sudden changes in pH (White, 1987) and maintaining optimum reactivity conditions for soil life.

**In biological properties:**

a) They encourage mineralization processes, thus contributing nutrients and energy to microbial life.
b) It facilitates the reactivity and absorption mechanisms of dangerous substances such as plaguicides and other toxic agents, helping their degradation (Harrod et al., 1991).
c) It can stimulate plant growth through the presence of substances that activate their physiological mechanisms and the control of diseases and pests (Vaidyanathan and Eagle, 1991), as it helps to maintain a balanced ecological system.

Cattle waste is formed by the accumulation of solid and liquid waste on cattle farms. The efficient use of this waste without damaging the environment, especially the liquids, is a priority objective for many researchers (Bernal and Roig, 1993).

**Use of organic supplements (INCA, 1998):**

When it is necessary and possible to use organic fertilizers to benefit soils and crops, before deciding what to do and how, some basic criteria should be borne in mind.

- That the beneficial effect of organic fertilizers on the soil and crops is determined by the amount of organic matter applied and not by the amount of organic fertilizer itself.

- That the chemical characterization of organic matter is essential in order to make any decision and the corresponding calculations.
- That the application needs of organic allowances depend on the objectives of these applications.

It's important to bear in mind that when applying organic matter to improve soils, the optimum quantities depend on the texture of the soil. Clay soils need higher doses, and the decomposition process is slower because in these soils there are more micro pores and less ire available for microorganisms. The decomposition time is longer and the residual effect lasts longer.

Sandy soils require lower doses. The decomposition process is faster because there are more macro pores in the soil and more ire and microbial activity is greater. The residual effect lasts less time.

**For long-cycle crops**

**For sandy soils**

It has been determined (Paneque and González, 1985) that the optimum doses are in the range of 15 to 25 $t.ha^{-1}$ of "*pure organic matter*". In these cases, the residual effect of the benefits obtained in the soil can last from 2 to 3 years.

**For clay soils**

The optimum doses are in the range of 25 to 40 $t.ha^{-1}$ of "*pure organic matter*". In this case, the residual effect lasts 4 to 5 years.

**For short-cycle crops**

When organic fertilizers are applied to soils dedicated to short-cycle crops, the recommended quantities are low.

**For sandy soils**

5 to 7 $t.ha^{-1}$ of "pure organic matter" is recommended.

In short-cycle crops, the residual effect of the organic matter lasts less time because when the crops are stopped and the soil is prepared for the next cycle, the decomposition of the organic matter is violent and accelerates.

**Determining the quantities of organic allowance to be applied**

In order to determine the amounts of organic allowance that should be applied in each case, the following information must be available:

a) Crop cycle.
b) Soil texture.
c) Fresh organic matter content, % moisture and N, P2O5 and K2O of the available organic matter.
d) Dose of pure organic matter recommended for these conditions.

**Quality indicators for an organic allowance**

The quality indicators of an organic allowance are given by:

1. **Its humus content and quality**

Humus from the decomposition of organic waste with a low C/N ratio and low lignin content is looser and more friable, and its microbial flora is more favorable to the physical and biological conditions of the soils where it is applied, than when the organic fertilizer is obtained from plant waste rich in lignin with a very high C/N ratio.

An organic crop must have 50 % or more organic matter in dry matter. Below 50% organic matter is considered poor quality.

2. **Its nutrient content**

Organic fertilizers must have a balanced N, P, K, CA, Mg nutrient content, so that when they are used they improve soil fertility and benefit the nutritional status of plants. In addition, organic fertilizers must not contain substances that acidify or alkalinize the soil and could affect the normal development of crops.

3. **Its moisture content**

Moisture content is the element that most limits the use of organic fertilizers due to the cost involved in transporting and handling large quantities of water in organic fertilizers.

In general, organic fertilizers have a moisture content of between 60% and 80%, depending on the production process and the nature of the organic waste they contain, which always tends to retain water. However, anyone who produces and manages organic fertilizers should make sure that they have 60% moisture or less when they are used. Organic fertilizers that are too wet not only increase transportation costs, but are also difficult to manage, apply and distribute properly.

4. **Its C/N ratio**

The C/N ratio is one of the most important characteristics of an organic fertilizer. It depends on its value:

a) Its speed of decomposition when applied to soils.
b) The fixation and mineralization of nitrogen in the soil and the possibility of competition between soil microorganisms and plants for this element.
c) The utilization of C from organic matter and its conversion into humus in the soil.
d) Their physical, chemical and biological properties.

The speed with which organic fertilizers decompose depends on how long you have to wait after applying them before sowing or planting the crop that will benefit from them.

Whenever organic fertilizers are applied to soils, microbial activity is encouraged because organic matter is a source of energy for microorganisms. In order for them to reproduce, develop and grow, they must take N, P, K and other nutrients from the environment. If these nutrients are not found in the organic matter, the microorganisms take them from the soil and then competence can be established between microbial activity and the development of the plants with which they coexist. Many authors have reported (Russell, 1967; Gros, 1966 and others) that the greatest competence between microorganisms and plants is produced by nitrogen.

**Evaluation of organic allowances**

In order to evaluate organic fertilizers, it is necessary to carry out a chemical analysis that characterizes them. The most important determinations are: Moisture, dry matter, pH, free carbonates, organic matter and the determination of total elements: N, P2O5, K2O, CA and Mg and calculating the C/N ratio.

This analysis can be carried out using the analytical methods of the A. O. A. C (1950) or the techniques described in the manual of analytical techniques of INCA's Agrochemistry laboratory.

The quality of organic fertilizers depends on many factors that are closely related to the origin and nature of the waste used in their composition, the fermentation process used and the chemical products used to enrich them. For this reason, the quality of the same organic fertilizer can vary from place to place, but there are indicators that are basic for evaluating it, regardless of its origin or source.

Requirements for organic allowances to be applied:

A. Humidity

- ✓ The lower the allowance, the higher the quality.
- ✓ It is desirable to have 60% or less.

B. C/N ratio

- ✓ It is desirable for the C/N ratio of organic fertilizers to be less than 25:1 for the reasons explained above.

C. Organic matter content

- ✓ Organic matter content is the basis of all organic fertilizer. Its content must be 50 % or more, expressed as dry matter.

d. Mineral nutrient content

- ✓ Although the mineral nutrient content is not the most important aspect in organic fertilizers to define their quality, their content has great economic and practical value, especially when establishing sustainable farming systems.
- ✓ It is desirable that the N, P2O5 and K2O content of organic fertilizers is balanced and as high as possible, so that when a given dose is applied, the mineral nutrients it contains are sufficient for the development of any crop, without the need for corrections through the application of chemical fertilizers.

Important aspects in preparing mixtures for substrates. Their use and exploitation in organoponics or other forms of organic farming. In order to achieve maximum efficiency in organoponics and obtain good crop development and the expected yields, some basic requirements for these growing conditions must be met.

The most important thing:

1- Ensure that the mixture has the right organic matter content for each specific condition.

2- That the balance of nutrients (N, P2O5 and K2O) is favorable for the crops and yields expected.

3- Prepare the mixes, making sure that the correct proportions are used according to the corresponding calculations. The substrates must have a uniform and homogeneous composition. Substrates are neither soil nor organic matter; they are integral compounds that have their own characteristics, which depend on the nature of the materials that make them up and the way in which they have been integrated (mixed) to obtain them.

In organoponics, the substrate is intensively exploited and microbial life is very active; all of which is conducive to organic matter decomposing fairly quickly.

Although the work of incorporating and mixing organic fertilizers with the substrates of established quarry workers is somewhat difficult, it is essential that these mixtures are made with care and rigor. If you don't manage to mix them well, the organic matter added will have very little effect on the crops and the residual effect of the organic matter will be very limited.

## 2.12 Farmyard manure

### Application of fresh manure

In ecological agriculture, surface dressing is practiced with great success. This consists of applying a thin, shallow layer of fresh manure over the farmland and spraying it on evenly. It is not advisable to incorporate the manure too deeply with the plow; it is preferable to leave it on the surface, as incorporating it too deeply does not decompose, it rots and affects the soil.

Fresh manure has a rapid and obvious effect on plant growth because of its high nitrogen content. Ecological agriculture does not seek accelerated growth, but harmonious and uniform growth, which can be guaranteed by healthy soil. It is important that the manure feeds the soil microorganisms first and not the plants directly. Healthy soils with adequate moisture assimilate manure in 2 to 3 weeks. However, surface applications of fresh manure can be washed away by run-off during irrigation or by rainfall (Kalmans and Vázquez, 1999).

### Dung tea

Manure tea is a preparation that converts solid manure into liquid manure. In the process of making tea, the manure releases its nutrients into the water, making them available to the plants. In the preparation of manure tea, 25 lbs are added to a bag of any type of manure, a large stone is added (to give it weight), the bag is tied tightly with a rope and then the bag is placed in a tank with a capacity of 200 L of water, covered and left to ferment for two weeks. After this time, the bag is removed and the manure tea is prepared (Manual do A. O. 2002).

To apply this fertilizer, dilute one part of manure tea in one part of fresh water and prune, then apply in bands to the crops or around the fruit trees as far as the branches extend. This fertilizer can also be applied through the irrigation line by distillation (200 L/ha) every 15 days (INIFAT, 2002).

## III. MATERIALS AND METHODS

This work was carried out on the property belonging to farmer Mr. Henrique Mariano, located behind the SONANGOL fuel pumps, in the town of Homaengue, Kuanhama municipality, Ondjiva Province. The experiment was carried out on 5 plots, which had an area of 1.50m x 25.0m. A randomized block experimental design was used. The time of sowing, transplanting and harvesting was between March 2023 and June 2023, according to the National Institute of Meteorology and Geophysics. Cunene Provincial Meteorology Department (Annex 1).

To apply the different doses of cattle manure, the plots were divided into 5 treatments, starting with a control and applying the doses listed below: 25%, 50%, 75% and 100%. The soil was moved with a fork and softened to a depth of 25 to 30 cm. The soil was then extracted according to the doses of cattle manure to be applied. For the mixtures of cattle manure and soil, the amount of manure and soil was measured with a wheelbarrow and crossed over several times with a shovel to ensure homogeneity in the treatments.

After the bovine manure had been mixed with the soil, the treatments were filled in with a coarse and then a fine sod until all the plots were well prepared. Sowing was carried out using the direct sowing method on March 21, 2023, with germination starting 5 days after sowing. Irrigation was applied in accordance with the technical instruction manual for organoponics, intensive vegetable gardens and semi-protected crops (2007).

Watering was carried out using a hose from a chinpaca near the experiment area, watering daily for the first 8 days, then every two days until 60 days and every four days for the remaining 30 days due to the texture of the soil.

The work carried out in the experiment was done by hand. The cattle manure used was sourced from various nearby corrals and from the farmer himself where the experiment was carried out. The manure underwent the decomposition process naturally, being located in a high area where there is no waterlogging, no streams and being very dispersed, which favors the transformations that occurred with the presence of a dark color, with no characteristic aroma and coming from the excreta of cattle that consumed natural pasture.

The seed was bought by the farmer himself in a store where these products are sold. The seed complied with all the established parameters, in sealed packages and in optimum condition, with good varietal purity and high germination power. 750 g of seeds were used for all the plots at a rate of 150 g per treatment. The plots had the following measurements: 1.50 m x 25.0 m (37.5 $m^2$ = 0.019 ha x 5 plots = 0.095 ha as a total area).

The Detroit Dark Rede variety was planted at a distance of 15 cm from the camellón without a nose mark because it was planted in chorrillo. 30 days after sowing, the thinning began, eliminating the weakest and the rest to establish the 10 cm nose mark. The first measurement and leaf count took place on April 20, 2023, with 294 plants per treatment (5) as the standard. Due to their rapid growth, only 4 measurements were taken as the leaves broke off easily. The rest of the measurements were taken after the harvest, such as the thickness of the fleshy root and the weight of the measured roots per treatment, as well as the weight of the treatment per $m^2$ and the total weight of the treatment. The remaining measurements were taken every seven days.

Cultivation work was carried out systematically in order to avoid boils that could damage the crop and allow it to acquire all the nutrients in the soil without being able to develop physiologically. For phytosanitary work, sampling was carried out every 5 days to determine the presence of pests and diseases. As there were no affections, a preventive application of Nin was carried out, as it is a biological means that does not affect humans or the environment. repellent plants such as corn and oregano were used. The dose used is in accordance with the Integrated Management of Pests and Diseases of Aromatic, Medicinal and Condiment Plants (1999).

The foliage was measured using a 1m tape measure, which showed the greatest growth in the 75% and 100% cattle manure treatments respectively, the root was measured using a king's foot and the weight was weighed in kg. The means were processed by Duncan's test using the statistical package soware.

Materials used in the experiment:

Hoes. Tractor with trolley

Shovel tape measure

Fork King's foot

Wheelbarrow weighbridge

Coarse and fine achinho Seed

Some concepts that were taken into account when carrying out the experiment:

a) *Sowing time: Sowing* crops at the right time to match the requirements of each plant species is one of the factors that favors plant growth and development.

The experiment was planted on March 21, 2023 and harvested on June 24-25, 2023, which corresponds with the sowing season and harvest time, which is between 80 and 90 days. Application ã of an agronomic strategy that allows us to avoid attacks from pests and diseases, advancing or delaying the sowing season as appropriate without compromising yields. Taking advantage of rainfall and temperatures according to plant growth and development.

b) *Sowing* density: controlling the sowing density ensures that the plants find the optimum living space for their development. Populations higher than those required increase the competition between plants and increase the shade they provide each other, which favors the development of fungal diseases due to excess humidity. Fewer plants per area increases the presence of weeds and does not result in economic use of the land.

c) *Layering:* this is done when the seedbed is too dense. It consists of eliminating the excess plants and leaving the most vigorous and developed ones at the recommended distance. This should be done at the right time for each crop, with care, separating the plants to be eliminated to one side, pulling them out gently and pressing the soil around those that remain to avoid damaging or pulling them out.

d) *Pruning the plots: this* can be done with an azadon or by hand, eliminating all the weeds between the plots, also to prevent pathogens from developing. This work is carried out when the weeds are small, these weeds are collected and deposited outside the area of the experiment. The experimental design was completely randomized and the averages were compared using Duncan's test using the statistical package soware.

**Cultural attention:**

The cultural care given to the crop was the following:

- ✓ Choice of posture.
- ✓ Removal of bad weeds from the plots every 7 days.
- ✓ Aporque.

- ✓ Daily observations to diagnose attacks by pests and diseases.
- ✓ Water irrigation.
- ✓ Application of preventive biological products.
- ✓ Harvest at 90 days.

## IV. RESULTS AND DISCUSSION

Each of the variables analyzed had the following results:

### 4.1 Length of foliage

This table shows the analysis of the foliage results at 30 and 51 days respectively, where it can be seen that there was no significant difference between the treatments. In both cases, the lowest value is occupied by treatment number 1 (control) and the highest value by treatment number 5 with a percentage of bovine manure of 100%.

**Table 8.** Foliage length (cm.)

| Treatments | Foliage length (cm.) | | | | | |
|---|---|---|---|---|---|---|
| | At 30 days. | | | At 51 days | | |
| | X | Meaning. | E. S | X | Meaning. | E. S. |
| T1 (witness) | 23.06 | N. S | 0.53227 | 37.18 | N. S | 0.9232 |
| T2 (m.o al 25 %) | 23.08 | | | 39.46 | | |
| T3 (m.o al 50 %) | 23.08 | | | 38.80 | | |
| T4 (m.o al 75 %) | 24.06 | | | 39.10 | | |
| T5 (m.o al 100 %) | 24.44 | | | 40.58 | | |
| CV = 5.055259 % | | | | CV = 5.290483 % | | |
| $P \leq 0.05$ | | | | | | |

It can be seen that the treatment with the 100 % manure percentage occupied the top spot, because this treatment contains a greater amount of nutrients, which stimulates the growth of the foliage and coincides with Manjaiah *et al.* (2000), who state that the release of exudates by the roots, as well as contributing organic matter, are important sources of energy for microorganisms and, therefore, greater plant development.

Theurer and Daney (1980) state that during the first few months of growth, the leaf apparatus dominates over the root apparatus and the infrastructure needed to produce sugar is consolidated.

According to Guenkov (1989), plant nutrition is very important for the formation of organs and the performance of all biological and physiological processes. In order to grow and develop, plants must not suffer any shortage of nutritional substances. Lighting is very important for this crop, as it allows photosynthesis to take place properly and conditions the importance of sugar production (Infoagro 2009).

According to Cooke (1987), any substance that is added to the soil to contribute one or more nutrients to the plant in order to increase its growth is a fertilizer. The importance of manure lies not only in the amount of organic matter it contains, but also in the nutritional principles it provides to plants (Alonso, 1980).

According to Vaidyanathan and Eagle (1991), organic fertilizers can stimulate plant growth through the presence of substances that activate the plant's physiological mechanisms and the control of pests and diseases. The growth of beet is adequate when the pH in the soil is 6 - 7. In acidic soils with a pH below 5 or alkaline above 8, the young plants show different physiological disorders, the seedlings sprout slowly and a large percentage of them perish (Infoagro, 2009).

**4.2 Number of leaves**

The size of the leaves depends on the nutrient area. The larger the nutrient area, the more developed the leaf system. Table 9 shows the results of the analysis of the number of leaves at 30 and 51 days. As a result, treatment number 1, corresponding to the control, had the lowest value and treatment number 5, with a percentage of 100% cow manure, had the highest value. There was no significant difference between the treatments because the percentages of manure had no influence on the number of leaves.

**Table 9.** Number of leaves (u).

| Treatments | Number of leaves (U) | | | | | |
|---|---|---|---|---|---|---|
| | 30 days. | | | At 51 days | | |
| | X | Meaning. | E. S | X | Meaning. | E.S. |
| T1 (witness) | 6.80 | N. S | 0.20000 | 9.40 | N. S | 0.3286 |
| T2 (m.o al 25 %) | 6.80 | | | 9.00 | | |
| T3 (m.o al 50 %) | 6.80 | | | 9.20 | | |
| T4 (m.o al 75 %) | 6.80 | | | 9.00 | | |
| T5 (m.o al 100 %) | 7.20 | | | 8.80 | | |
| CV = 6.500198 % | | | | CV = 8.093028 % | | |
| $P \leq 0.05$ | | | | | | |

According to Hernández, (2007) the right amounts of nitrogen are fundamental in the formation of beet foliage. In order to produce larger fruit and in greater quantity, plants must first form a well-developed leaf system. This can be achieved with sufficient nutrients during the period when the leaf system is being formed (Guenkov, 1989).

According to Guenkov (1974), cow manure contains substances that stimulate branching. Studies carried out by INSAM (1996) show that there is a clear analogy with the effects on plant development of the application of humic substances on the mineral nutrition of plants, with effects on the accumulation of phosphorus, nitrogen, potassium, calcium and magnesium. The development of crops is based on the soil's ability to provide the necessary quantities of nutrients for their correct development.

According to Infoagro (2009), excess nitrogen increases leaf development, but decreases the ability to mobilize sugars to the root.

**4.3 Di â meter of the bulb (cm.)**

In order to evaluate Table 10 (bulb diameter), the bulbs were measured after harvesting; the lowest value was taken by the control treatment and the highest by treatment number 5 with a percentage of cow manure of 100%. It can be seen that there was no significant difference between the treatments, so it can be concluded that cow manure had no influence on bulb diameter.

**Table 10.** Bulb diameter (cm.)

| Treatments | X | Meaning | E. S |
|---|---|---|---|
| T1 (witness) | 6.42 | N. S | 0.135204 |
| T2 (m.o al 25 %) | 6.48 | | |
| T3 (m.o al 50 %) | 6.62 | | |
| T4 (m.o al 75 %) | 6.84 | | |
| T5 (m.o al 100 %) | 6.86 | | |
| CV: 4.550336 % | | P < 0.05 | |

According to Infoagro (2009), in order to get a good yield from this crop, it is necessary to drill 35-45 cm to facilitate good root development.

According to Guenkov (1981), the diameter of the beet for harvesting should be between 6-8 cm, which corresponds with the data from the experiment shown in the table. Deep soils with a pH of around 7, high water retention capacity, low crust density and good aeration are the most suitable for growing beet (Infoagro, 2009).

According to Infoagro (2009), beet can be sown on different soils, as long as they provide an adequate balance of nutrients and moisture. The most suitable are deep soils with good texture and absorption capacity and permeable soils with a high humus content.

Beet does not develop fully in any situation, as it is very demanding on the soil, since the growth of the root and its quality are directly influenced by the quality of the soil (Orange, 2009).

**4.4 Bulb weight (Kg.)**

Table 11 shows the statistical analysis of bulb weight, in which there was a significant difference between the treatments, so it can be concluded that cattle manure has an influence on bulb weight. The highest value was found in treatment number 5 with a percentage of cattle manure of 100%, which showed a significant difference with treatment number 4, and these also showed significant differences with the rest of the treatments, since the lowest values were found in the treatments with the lowest percentage of cattle manure.

**Table 11.** Bulb weight (Kg.)

| Treatments | X | E. S |
|---|---|---|
| T1 (witness) | 20.74 a | 0.250679 |
| T2 (m.o al 25 %) | 20.08 a | |
| T3 (m.o al 50 %) | 20.08 a | |
| T4 (m.o al 75 %) | 21.72 b | |
| T5 (m.o al 100 %) | 22.68 c | |
| CV: 2.661612 % | | P < 0.05 |

According to Theurer and Doney (1980), early foliage development in spring is always linked to high root weight productivity. Azzan and Samuel (1964) cited by Garandilla, (1991) state that for vegetables, it has been observed that applications of organic fertilizers have had beneficial effects on yields.

Infoagro, (2009) states that climate is one of the factors that directly affects yields. A temperate, sunny and humid climate contributes to the production of a high percentage of sugars in sugar beet.

Gorinij (1961) and Jaibebo, (1967) agree that mineral fertilizer increases the harvest more than manure, only during the first few years, but that in subsequent years the harvest is higher with manure. Beet prefers arcillo-limestone soils that are airy and cool, rich in well-decomposed organic matter and potassium, with a pH of between 6 and 7 (Infoagro 2009).

The soil must contain enough of the fundamental nutrients throughout the plant's growing cycle (Guenkov, 1974). It is important to balance the nutrients in the soil well in order to obtain good quality fleshy roots. Water is the factor that most influences the weight and richness of beet; it is also the most difficult to manage, as it depends on many other parameters such as climate, soil type and root depth (Infoagro, 2009).

Cavalheiro and *et. al*. (1998) state that with the application of 10.0 Kg/m$^2$ of cattle manure they were able to increase yields almost twice as much as the control, as this material produces changes in physical and chemical properties.

**4.5 Economic Value**

Table 12 shows the results of the economic value of each treatment, or the indicators that helped us make decisions about applying one or other dose of cattle manure. To do this, we started with the yields, which are the indicators that support us in deciding whether or not to apply a given dose.

It can be seen that the highest yields were achieved by the treatment in which 100% cattle manure was applied, followed by the treatment with 75%, and the lowest values were achieved by the treatments with 25% and 50% cattle manure respectively.

The treatment with 100% bovine manure was the best performing with a yield of 37,793 t.ha$^{-1}$ and a gain in realization of Akz 24 639.00, a clear gain of Akz 6 515.00 and a cost per weight of AKz 0.73. Despite the fact that this treatment has the highest expenditure on cattle manure, it has a productive backing due to the positive effect of cattle manure on beet cultivation.

**Table 12.** Economic value

| Treatments | Yield (t.ha$^{-1}$) | Realization profits (AKz) | Expenditure on organic material (AKz) | Other expenses (AKz) | Total Expenses (AKz) | Sharp Gain (AKz) | Costs per Pesos (AKz) |
|---|---|---|---|---|---|---|---|
| T1 (witness) | 34.560 | 22570.00 | - | 9856.00 | 9856.00 | 12714.00 | 0.43 |
| T2 (m.o al 25 %) | 33.460 | 21834.00 | 2130.00 | 9856.00 | 11986.00 | 9848.00 | 0.55 |
| T3 (m.o al 50 %) | 33.460 | 21834.00 | 3635.00 | 9856.00 | 13491.00 | 8343.00 | 0.61 |
| T4 (m.o al 75 %) | 36.192 | 23600.00 | 4740.00 | 9856.00 | 14596.00 | 9004.00 | 0.61 |
| T5 (m.o al 100 %) | **37.793** | **24639.00** | **8268.00** | **9856.00** | **18124.00** | **6515.00** | **0.73** |

| total | 175.465 | 114477.00 | 18773.00 | 49280.00 | 68053.00 | 46424.00 | 0.59 |
|---|---|---|---|---|---|---|---|

## V. CONCLUSIONS

1) The variable with 100% cattle manure had the highest values for bulb weight.
2) The percentages of cattle manure had no influence on foliage, bulb diameter and leaf length.

## VI. RECOMMENDATIONS

1) Apply high percentages of cattle manure to beet crops.
2) Carry out the experiment in other production units with cattle manure under other climatic conditions

## VII. BIBLIOGRAPHY

Alonso. F. (1980). Compendio de suelos. Ed. Pueblo Educación instituto cubano de libro La Habana 134-151p.

Altieri, M. (1989). The Transition from Conventional Agriculture to Ecological Agriculture. Revista Cultivando (9): 19-25, August.

Altieri, M. (1991). Traditional farming in Latin America. The Ecologist 21: 93-96.

Alcántara, A. F. (1993). Residuos agrícolas, forestales, ganaderos e industriales. Ed. Instituto de investigaciones ecológicas. Málaga.

ACTAF, (2008). VII Encuentro de la Agricultura Orgánica y Sostenible, Año 14 N 2.

Bernal, M. P. and Roig, A. (1993). Nitrogen Transformations in calcareous soils amendedwith pig slurry under aerobit incubation. J. Agric. Sci, 120, 89-97.

Cegarra, J., Hernández, t. Y Costa, F. (1983). Adición de residuos vegetales a suelos calizos. V. influencia sobre el desarrollo vegetal. An. Edafol. Agrabiol. 42 (3-4), 545-552. 8-

Cooke. W. G. (1987). Fertilizantes y sus usos Ed. Continental, S.A. Dec. U. México 46p

Christensen, B. T. (1996). Matching measurable soil organic matter fractions with conceptual pools in simulation models of carbon turnover: Revision of model structure. P.144-160. *In* Powlson D. S., P. Smith, and J. Smith (Eds). Evaluation of soil organic matter models using long-term datasets. U. NATO ASI Series I: Global Environmental Change. Vol I 38. Springer- Verlag, Berlin, Heidelberg, Germany.

State Statistics Committee (CEE) (1997). Statistical Yearbook of Cuba.

Caballero. R. *et al.* (1998). Uso de humus de lombriz en la fertilización de las hortalizas en huertos intensivos XI Seminario Científico del INCA, Programa y Resúmenes p-21

Canet, R., R. Albiach, F. Pomares (2000). Biological activity indices as a diagnostic tool for soil fertility in ecological agriculture. *En Investigación y perspectivas de la enzimología de suelos en España* (C. García, M.T. Hernández eds.) CSIC-CEBAS Murcia. pp. 11-39.

Cuba, Ministry of Agriculture (2000). Manual técnico de organopónicos y huertos Intensivos. La Havana: Editorial AGRINFOR, 55 p.

Dinchev. T. (1977). Agrochemistry. E d. y Educación la HABANA. p -253

Dick, R. P. (1994). Soil enzyme activities as indicators of soil quality. In *defining soil quality for a sustainable environment* (J. W. Doran *et al.* eds). SSSA special publication n. 35 SSSA and ASA, Madison. pp. 107-124.

Diez, J, A. (1997). Soil pH dynamics affected by the application of different organic fertilizers. An. Edafol. Agrobiol. 46 (3-4), 499-510.

University of Illinois Extension (2009). Watch your garden grow. Retrieved May 3, 2022 from http: // WWW. Vea su jardín crecer.com.

Ferruzzi, C. (1987). Manual de lombricultura. Mundi Prensa. Madrid: Spain.

Gorinij. C. (1961). Influence of continuous applications of Fumier gor IA cumulation of soil hums on the yield of crop plants 86 -93 p.

Guenkov G. (1969). Fundamento de la horticultura cubana. ED. Pueblo y educación.

Guenkov, Guenko (1989). Fundamentals of Cuban horticulture. La Habana: Cuban Book Institute: Editorial Pueblo y Educación, 1974. 458 p. 22-

Garay, J. (1979). De donde venimos. Recorrido histórico desde el origen del hombre hasta el triunfo del capitalismo. Second edition. Venezuela. 58

Gabriels, D. and Michiels, p. (1991). Soil organic matter and water erosion processes. Advances in soil organic matter research: the impact on agriculture and the environment >>, 141-152. Ed. The Royal Society of chemistry, Cambridge (United Kingdom).

García F. (1992). Study on the stability of aggregates in the soil, chemical and microbiological aspects. Degree thesis from the Faculty of Biological Sciences of the University of Murcia. Murcia.

Golchin, J., M. Oades, J.O. Skemstad, and P. Clarke (1994). Soil structure and carbon cycling. Aust. J. Soil Res. 32:1043-1068.

Harrod, T. R. Carter, A.D. y Holles, J. M. (1991). the role of soil organic Motter in pesticide movement via run-off; soil erosion and leaching. Advances in soil organic matter research: the impact on agriculture and the environment>>, 127-138. Ed. The royal society of chemistry, Cambridge (United Kingdom).

Huerres. C. Y Caballero. B. (1996). Horticultura E d. Pueblo y Educación La Habana. 21p

Gros, A. (1996) Abonos. Practical guide to fertilization. Edición Revolucionaria. La Habana.

Gómez and H Laterrot. (2002). Manual técnico de organopónicos y huertos intensivos.

García, (2007). Study of the cultivation of Remolacha (Beta Vulgaris L.). Under intensive Huerto conditions. Course work.

Hernández, V. G. (2007). Livestock and agricultural activities. Prontuario. Third edition (ACPA). 46-47.

INSAN (1996). Microorganism and Humus in soil En Humic substances in terrestrical ecosystems. Piccolo. A. Ed Elsevier A Amsterdam 265 -292 p

INCA (1998). Organic subsidies. Conceptos prácticos para su evaluación y aplicación. Victor and Pérez P. pp5-28

INIFAT (2002). Manual de Abonos Orgánicos para la Agricultura urbana en Cuba. Ramírez C. and Martínez F. pp 7-8

Jaivebo (1967). Influence of Fertilizer and monore additions en monex chanageable ammonium soil sei 16-28p.

Juma, N. G., W. B. McGill (1986). Decomposition and nutrient cycling in agroecosystems. En *Micro floral and faunal interactions in natural and agro ecosystems* (M. J. Mitchell, J. P. Nakas Ed.) Dordrecht. pp. 74-136.

Kalmans, E. Y Vásques, D. (1999). Manual de agricultura ecológica. Second edition. 63-64.

López. N. y Zada. N (1989). El Plátano, Editorial Pueblo y Educación La Habana 141 -146 p.

Labrador, J. (1996). *Organic matter in agricultural systems*. MAPA and Mundiprensa, Madrid.

Technical Manual for Organoponic, Intensive Huerto and Semi-protected crops and IIRD (2007). Sixth edition, Ciudad de la Habana.

Nuñez, M. (1994). Ecological and Sustainable Fruit Growing in the Tropics. --Mimeography. --Mérida, Venezuela: IPIAT Publications. -- 24p.

Ocsa, Walter (1995). The walipini system. Published by CEFODCA. La Paz, Bolivia. 36 pp.

Pérez. N. y Nilda. Caraballo (1968). Manejo agro ecológico de plagas, en agro ecología y agricultura sostenible. Modulo 2, ISCAH Habana.

Paneque, V. M. and P. J. González (1985). Evaluación de la cachaza como abono químico para la caña de azúcar cultivado en suelo Ferra lítico Cuarcítico. Memorias de la III Jornada Científica del Instituto de Suelo. La Habana.

Pineda, R. (1994). Lombricultura. Humus de lombriz: preparación y uso. CIPCA-PIURA: Peru.

Russell, E. W. Russell (1967). Soil conditions and plant development. Editora Revolucionaria. La Habana.

Riveiro, D. (1973). El proceso civilizatorio. Edición de la Universidad Central de Venezuela. Caracas. Venezuela. 196p

Resolución Económico del V Congreso del PCC. (1997). La Habana. Cuba.

Redino, C. and Pedraza, M. (2002). Centro de Estudio para el Desarrollo de la Producción Animal CEDEPA. E. Mail redi a reduc. Cmw.edu.cu.

Rodríguez Nodal et al Collaborators (2007). Technical Manual for Organoponics, Intensive Huertos and Semi-protected Organoponics ACTAF.

Magazine: Vea su jardin crecer. extracted on January 25, 2021 from http://www.Remolacha.com.

Beta Vulgaris Magazine. Retrieved February 21, 2022 from http:// www.Wikipedia, la enciclopedia libre.com

Ecological agriculture magazine. Retrieved February 27, 2022 from http://www.Agrícultura.com.

Magazine Concept of pH and importance in Fertirrigation. Retrieved March 20, 2022 from http://. WWW. Copyright infoagro.com.

Remolacha azucarera cultivation magazine (1st part). Extracted on March 25, 2022 from http: // WWW. Copyright infoagro.Com.

Revista El cultivo de la remolacha azucarera (section 1 to 2). Extracted on June 22, 2022 from http: // WWW. Copyright infoagro.com.

Swifter. S. (1991). Effects of Humic substances and polysaccharides on soil aggregation. En << Advances in organic matter research: the impact on agriculture and the environment >>, 153- 62. Ed. The Royal Society of chemistry, Cambridge (United Kingdom).

Turco, R. F., A. C. Kennedy, M. D. Jawson (1994). Microbial indicators of soil quality. In *defining soil quality for a sustainable environment* (J.W. Doran *et al.* eds.). SSSA special publication n. 35 SSSA and ASA, Madison. pp. 73-90.

Turruella, P. E., Ramírez, C. M., Martínez, F. (2002). Manual de abonos orgánicos para la Agricultura urbana en Cuba. (I N I F A T. 2002). 71.

Vaidyanathan, L. V. and Eagle, D. J. (1991). The influence of organic matter and clay on adsorption of atrazin by top soils. En<<advances in organic matter research: the impact on agriculture and the invironment >>, 381-391. Ed. The Royal Society of chemistry, Cambridge (United Kingdom).

Whithe, R. E. (1987). Introduction to the principles and practice of soil science. Ed. Blackwell Scientific publications, Oxford (United Kingdom).

Waters, A. G. and Oades, J. M. (1991). Organic Motter in waters table aggregates. In <<Advances in soil organic Motter research: the impact on agriculture and the environment >>, 299-314. Ed. The Royal Society of chemistry, Cambridge (United Kingdom).

# VIII. ANNEXES

## Annexes 1. climatic variables for the period

MINISTÉRIO DAS TELECOMUNICAÇÕES, TECNOLOGIA DE INFORMAÇÃO E COMUNICAÇÃO SOCIAL
INSTITUTO NACIONAL DE METEOROLOGIA E GEOFISICA
DEPARTAMENTO PROVINCIAL DE METEOROLOGIA DO CUNENE

**MAPA DE TEMPERATURAS MÁXIMA E MINIMA, HUMIDADE RELATIVA E PRECIPITAÇÃO REFERENTE AO I TRIMESTRE DE 2023 – DADOS CLIMATICOS**

| Dias | Tª Máxima °c | | | Tª Mínima °c | | | Humidade Relativa % | | | Precipitação mm | | |
|---|---|---|---|---|---|---|---|---|---|---|---|---|
| | Jan | Fev | Marc | Jan | Fev | Marc | Jan | Fev | Marc | Jan | Fev | Marc |
| 01 | 30,3 | 29,2 | 33,1 | 19,3 | 20,0 | 19,7 | 70 | 61 | 51 | | | |
| 02 | 30,4 | 30,9 | 29,8 | 21,0 | 21,5 | 19,1 | 61 | 51 | 66 | | | |
| 03 | 23,7 | 30,1 | 32,8 | 19,8 | 21,3 | 19,8 | 67 | 67 | 44 | | | |
| 04 | 20,1 | 32,1 | 33,0 | 18,7 | 20,8 | 18,0 | 98 | 48 | 20 | | | |
| 05 | 29,8 | 31,4 | 34,0 | 19,0 | 22,2 | 18,4 | 40 | 30 | 22 | | | |
| 06 | 31,1 | 30,9 | 35,9 | 19,6 | 20,8 | 18,8 | 29 | 59 | 21 | | | |
| 07 | 33,7 | 26,7 | 35,9 | 20,8 | 20,9 | 21,4 | 31 | 58 | 35 | | | |
| 08 | 32,3 | 28,6 | 31,0 | 22,9 | 21,7 | 21,5 | 27 | 65 | 50 | | | |
| 09 | 31,3 | 31,7 | 34,7 | 21,2 | 19,4 | 22,6 | 30 | 43 | 30 | | | |
| 10 | 33,2 | 33,1 | 34,4 | 23,1 | 18,8 | 22,4 | 44 | 46 | 40 | | | |
| 11 | 28,9 | 33,1 | 32,0 | 21,2 | 20,0 | 20,8 | 75 | 31 | 78 | | | |
| 12 | 31,6 | 33,8 | 32,8 | 21,6 | 20,3 | 22,6 | 67 | 31 | 68 | | | |
| 13 | 30,7 | 35,1 | 34,8 | 19,9 | 21,7 | 21,0 | 62 | 35 | 40 | | | |
| 14 | 31,1 | 35,6 | 35,0 | 18,0 | 23,4 | 21,2 | 68 | 31 | 39 | | | |
| 15 | 31,9 | 35,2 | 34,3 | 18,0 | 21,6 | 21,1 | 50 | 22 | 33 | | | |
| 16 | 29,6 | 33,3 | 33,4 | 20,7 | 18,2 | 20,2 | 71 | 27 | 47 | | | |
| 17 | 26,4 | 32,9 | 31,7 | 20,0 | 16,9 | 17,7 | 67 | 31 | 40 | | | |
| 18 | 25,7 | 31,9 | 33,5 | 19,0 | 13,7 | 20,0 | 85 | 23 | 50 | | | |
| 19 | 25,1 | 33,0 | 33,3 | 18,1 | 15,0 | 19,4 | 96 | 22 | 46 | | | |
| 20 | 22,3 | 33,1 | 33,9 | 20,6 | 15,4 | 19,3 | 90 | 24 | 49 | | | |
| 21 | 27,2 | 34,2 | 32,6 | 18,6 | 14,8 | 20,2 | 78 | 22 | 55 | | | |
| 22 | 24,6 | 35,4 | 32,4 | 17,2 | 17,4 | 21,3 | 71 | 26 | 47 | | | |
| 23 | 27,4 | 34,0 | 29,8 | 19,4 | 21,3 | 18,8 | 79 | 39 | 72 | | | |
| 24 | 28,4 | 32,9 | 32,0 | 18,5 | 20,8 | 20,5 | 79 | 48 | 53 | | | |
| 25 | 30,2 | 34,5 | 30,8 | 20,6 | 20,5 | 18,5 | 68 | 49 | 46 | | | |
| 26 | 31,0 | 34,3 | 25,6 | 19,0 | 20,2 | 21,2 | 71 | 46 | 84 | | | |
| 27 | 31,4 | 32,5 | 30,6 | 20,2 | 21,6 | 18,3 | 70 | 64 | 55 | | | |
| 28 | 30,1 | 27,9 | 28,4 | 21,1 | 19,7 | 18,6 | 68 | - | 77 | | | |
| 29 | 30,3 | | 31,0 | 20,1 | | 20,3 | 59 | | 66 | | | |
| 30 | 27,8 | | 31,8 | 17,9 | | 20,6 | 77 | | 63 | | | |
| 31 | 27,4 | | | 19,6 | | 18,9 | 76 | | 62 | | | |
| Total | 901,6 | 906,3 | 1006,6 | 548,8 | 548,8 | 622,2 | | 1099,0 | 1565,0 | | | |
| Média | 28.6 | 30,2 | 32,5 | 19,7 | 19,6 | 20,1 | 65 | 78 | 51 | | | |

MINISTÉRIO DAS TELECOMUNICAÇÕES, TECNOLOGIA DE INFORMAÇÃO E COMUNICAÇÃO SOCIAL
INSTITUTO NACIONAL DE METEOROLOGIA E GEOFISICA
DEPARTAMENTO PROVINCIAL DE METEOROLOGIA DO CUNENE

**MAPA DE TEMPERATURAS MÁXIMA E MINIMA, HUMIDADE RELATIVA E PRECIPITAÇÃO REFERENTE AO II TRIMESTRE DE 2023 – DADOS CLIMATICOS**

| Dias | Tª Máxima °c | | | Tª Mínima °c | | | Humidade Relativa % | | | Precipitação mm | | |
|---|---|---|---|---|---|---|---|---|---|---|---|---|
| | Abril | Maio | Junho | Abril | Maio | Junho | Abril | Maio | Junho | Abril | Maio | Junho |
| 01 | 31,1 | 33,0 | 31,5 | 20,2 | 16,7 | 13,8 | 61 | 30 | 25 | | | |
| 02 | 32,0 | 32,5 | 32,1 | 19,4 | 15,1 | 15,5 | 59 | 27 | 29 | | | |
| 03 | 32,2 | 33,0 | 31,3 | 20,6 | 14,4 | 13,0 | 50 | 30 | 27 | | | |
| 04 | 31,8 | 32,5 | 32,7 | 20,6 | 14,4 | 12,4 | 55 | 31 | 22 | | | |
| 05 | 31,2 | 32,3 | 31,8 | 18,8 | 16,5 | 11,4 | 43 | 30 | 21 | | | |
| 06 | 32,6 | 32,6 | 31,8 | 17,8 | 16,5 | 11,4 | 54 | 25 | 13 | | | |
| 07 | 33,8 | 33,1 | 31,6 | 18,0 | 15,2 | 13,4 | 45 | 31 | 17 | | | |
| 08 | 33,4 | 32,8 | 30,9 | 19,6 | 17,8 | 12,6 | 39 | 35 | 21 | | | |
| 09 | 32,6 | 32,8 | 30,4 | 20,3 | 16,3 | 13,4 | 42 | 40 | 19 | | | |
| 10 | 33,6 | 33,5 | 31,2 | 21,3 | 16,3 | 13,4 | 46 | 24 | 20 | | | |
| 11 | 33,9 | 32,8 | 30,7 | 19,5 | 18,3 | 11,4 | 34 | 26 | 20 | | | |
| 12 | 33,4 | 31,7 | 27,2 | 19,5 | 15,1 | 11,8 | 40 | 30 | 22 | | | |
| 13 | 34,4 | 31,8 | 29,2 | 18,0 | 15,9 | 7,8 | 39 | 37 | 17 | | | |
| 14 | 33,7 | 31,7 | 29,4 | 19,8 | 14,2 | 9,2 | 35 | 28 | 24 | | | |
| 15 | 33,9 | 31,5 | 29,4 | 18,5 | 15,0 | 9,3 | 37 | 21 | 21 | | | |
| 16 | 33,4 | 30,2 | 30,1 | 16,6 | 19,9 | 7,7 | 33 | 34 | 15 | | | |
| 17 | 32,6 | 30,3 | 30,1 | 17,7 | 13,4 | 7,8 | 37 | 30 | 15 | | | |
| 18 | 32,7 | 30,5 | 30,0 | 16,7 | 11,2 | 9,2 | 45 | 24 | 18 | | | |
| 19 | 33,4 | 31,0 | 29,9 | 18,7 | 11,3 | 7,3 | 46 | 25 | 22 | | | |
| 20 | 31,6 | 32,3 | 30,0 | 20,2 | 12,3 | 8,1 | 45 | 18 | 22 | | | |
| 21 | 20,9 | 32,4 | 29,7 | 17,9 | 13,3 | 10,8 | 75 | 20 | 22 | | | |
| 22 | 29,2 | 33,1 | 29,6 | 18,0 | 15,6 | 10,8 | 72 | 25 | 20 | | | |
| 23 | 32,5 | 32,4 | 29,3 | 19,4 | 14,5 | 9,0 | 45 | 20 | 21 | | | |
| 24 | 32,7 | 32,4 | 28,3 | 20,6 | 16,4 | 11,2 | 46 | 28 | 22 | | | |
| 25 | 32,6 | 31,1 | 28,5 | 20,8 | 15,2 | 11,0 | 56 | 23 | 27 | | | |
| 26 | 33,3 | 31,7 | 28,5 | 19,1 | 13,4 | 9,5 | 48 | 21 | 34 | | | |
| 27 | 33,6 | | 23,0 | 18,2 | 15,1 | 9,4 | 39 | 23 | 52 | | | |
| 28 | 33,5 | 31,7 | 22,4 | 18,2 | | 3,3 | 28 | | 32 | | | |
| 29 | 33,7 | 31,9 | 27,1 | 16,4 | 15,9 | 3,6 | 35 | 20 | 21 | | | |
| 30 | 33,4 | 29,8 | 28,2 | 16,0 | 12,6 | 6,8 | 33 | 23 | 27 | | | |
| 31 | | 30,7 | | | 11,5 | | | 22 | | | | |
| Total | 973,6 | 959,1 | 885,9 | 566,4 | 449,3 | 305,3 | 1362 | 801,0 | 661,0 | | | |
| Média | 32,4 | 31,9 | 29,5 | 18,9 | 14,9 | 10,1 | 45 | 27 | 22 | | | |

Printed by Books on Demand GmbH, Norderstedt / Germany